GW01185932

SBN 11 290039 9

Cover: Dress Chariot 19th Century (see Plate 1)

Note: All the items illustrated in this booklet
are part of the Science Museum collections

A Science Museum
illustrated booklet

CARRIAGES

to the end of the nineteenth
century

by Philip Sumner, B.Sc.

Her Majesty's Stationery
Office London 1970

Introduction

Horse-drawn carriages, recognizable as such, were in use by the 16th century, having evolved from the primitive but elaborately decorated wooden cart-like vehicles of the mediaeval aristocracy. They had retained the ornate decoration and had greatly improved in comfort and in general design. A form of suspension had been introduced—the body was suspended by leather straps from vertical wooden supports at front and rear. Later, in an attempt to increase stability, a heavy wooden beam known as a 'perch' was used to join the two axles and in fact this remained a feature of certain types of design to the end of the carriage era. From early in the 17th century repeated efforts were made, with limited success, to find a satisfactory form of steel springing for the carriage body and these continued during the 18th century. The vertical supports which held the strap suspension were replaced first by upright laminated steel whipsprings, then by more curved forms known as elbow springs.

The tendency towards greater ostentation and luxury also continued during the 17th century and this, together with their increasing numbers, gave rise to some feeling against carriages. They were regarded by many as extravagant and a public nuisance. Taxes were imposed on them, both as a source of Royal revenue and as some deterrent to their use. Road taxes and amounts and types of duty on carriages varied from time to time, but they continued to be levied throughout the period of their history.

Nevertheless carriages continued to become more and more a symbol of status and prestige (and in fact this was so throughout the 18th and most of the 19th centuries). A great deal of money was spent on them. They could be drawn by two, four or six horses and were driven by a coachman or by postillions, i.e. a rider on each near-side horse.

Royal vehicles, in order to be readily distinguishable from those of the nobility, became extravagantly ornate, decorated with the Royal Insignia and with most elaborate carving and painting.

By the middle of the 17th century the making of carriages had become an art and in 1677 Charles II granted a Charter to the Worshipful Company of Coachmakers, who kept a watchful eye on the standard of craftmanship. No coachmaker could ply his trade without their permission and he had to be master of many skills and trades.

Despite the luxurious appearance of carriages, however, travelling was still a hazardous ordeal throughout the 17th century, except perhaps in the cities. Roads were usually in an appalling state, sometimes with ruts so deep as to make them almost impassable. Even when steel springs were used, the carriage bodies were subject to excessive jolting. The general framework was usually made of ash, since its elasticity helped to absorb shock and it was less liable to warp.

Carriages were still heavy and ponderous and their strength and durability was to some extent limited by the strength and design of the wheel. 'Dished' wheels therefore appeared about the middle of the 17th century—i.e. they were slightly cone-shaped with the end of the axle bent downwards so that as the wheel revolved the lower spokes were vertical. This type of wheel tended to remain more securely on the axle.

All the carriages mentioned so far were those of the wealthy, but meanwhile there had appeared in Europe, about the middle of the

16th century, a comparatively new type of vehicle. It was closed and design to carry 4 passengers inside, with luggage on the roof. It became known as a 'coach'—possibly from its town of origin, believed to be Kotze in Hungary. It was not designed for fare-paying passengers; but it may be regarded as the forerunner of the later well-known stage coach, which was in use in England by 1640 and which provided for the first time a new means of transport for the less well-to-do.

Increasing traffic on the roads led to continual complaints about their condition, but little was done about it, in spite of the fact that the General Highways Act of 1555 had empowered local authorities to compel parishioners to devote each year a certain number of days' work towards road repairs. In the 1660's the Turnpike Toll system was introduced, to extract payment from travellers for road maintenance, but again this resulted in little improvement. In 1745 however, a vital movement of cavalry and artillery was seriously delayed because of the amount of mud and the deep ruts. This caused rather more urgent attention to be given to the problem, as did the ever-increasing amount of traffic, to which had now been added many hired travelling carriages.

Towards the end of the 18th century roads were sufficiently improved to make possible the introduction of very much lighter owner-driven carriages, drawn by lighter, faster horses and for driving to begin to become a pleasure and a sport, as it had been for some time in Europe. Another result was the introduction of specially fast coaches to carry the mails, which before 1784 had been delivered by postboys on horseback. The number and speed of stage coaches, too, increased all over England.

Attention was also being paid to various technical improvements in the vehicles themselves. In 1767 J. Hunt invented a satisfactory method of shrinking a metal band tyre on to a wooden wheel rim. For a long time it had been obvious that improved methods were needed to retard a vehicle on a steep hill or hold it stationary when required. Patents for various devices for use on heavy vehicles began to appear early in the 17th century (although the lighter private carriages did not normally use brakes until the middle of the nineteenth century). At first the method usually adopted when descending steep hills was a chain secured skid-shoe on to which a stationary rear wheel rested: later wood or rubber blocks acting direct on the wheel rims were used, moved by hand lever or foot pedal or by a hand-wheel and screw.

It was during the early years of the 19th century that really effective improvements of all kinds were achieved. Two satisfactory methods of steel springing appeared—the C-spring, replacing the elbow spring (see photograph of Dress Chariot) and Elliott's elliptic spring of 1804, placed between the body and the axle (see photograph of Landau). John Macadam, appointed Surveyor-General of Roads in 1827, was responsible for an immense and widespread improvement in road surfaces. Thomas Telford cut through hills and built bridges. Private carriages increased enormously in variety, as well as in safety and, whether coachman-driven or owner-driven, they were no longer to be the prerogative of the wealthy: they were coming within reach of the professional classes.

The Royal Mail coaches attained a great reputation for speed, punctuality and reliability. By the 1830's there were more than 700 operating day and night, drawn by four horses and sometimes averaging more than 11 mph if the horses were changed about every 15 miles. There were also well over 3000 stage coaches in regular service, many of them equally fast and well-maintained.

Vehicles for use only in cities were also increasing in number and type, although, city roads being better, short journeys by carriage had been acceptable at an earlier date. In 1605 a few carriages for hire appeared—the hackney carriages—in London the first 'stand' was established in 1634. They were so successful that there were more than 3000 in London by the end of the 17th century, causing much resentment among the Thames watermen who had previously enjoyed a monopoly of London's passenger traffic. They were usually discarded private carriages and they continued in use throughout the 18th century.

In 1823 there appeared the first successful licensed two-wheeled vehicles for hire, the hackney cabriolets. In the 1830's, however, these and the hackney carriages were joined by and finally supplanted by the two-wheeled hansom cabs (actually designed by John Chapman in 1836 but based on a design produced two years earlier by Joseph Hansom) and the four-wheeled Clarence cabs (the 'Growlers'), both of which survived until the end of the 19th century. Another kind of horse-drawn vehicle appeared in London in 1829 and was immediately successful. This was the passenger omnibus, which brought travel within reach of the ordinary middle-class citizen and which also survived until replaced by a motorized version. Congestion on the roads became a real problem in large cities, with frequent chaotic traffic jams.

During the 20th century the motorized vehicle, bringing in due course its own traffic problems, replaced the horse-drawn vehicle, with one notable exception, the State Coach. This, and other elegant open carriages, have continued to be used by Royalty on ceremonial occasions to the present day.

In the following pages photographs of twenty representative Museum Carriages are shown, grouped according to their design and purpose.

1 Dress Chariot 19th century

In the first four pages some of the many types of coachman-driven vehicle of the 19th century are shown—from the Dress Chariot of the aristocrat to the small closed carriage for everyday use by the professional man.

The Dress Chariot was a nobleman's carriage used for ceremonial occasions, smaller but no less elegant than the Dress Coach. These luxuriously appointed vehicles were elaborately decorated in the family colours, with armorial bearings and crests. They carried only two passengers and were usually drawn by two horses. The coachman was in splendid livery and two footmen stood at the back, also in livery and carrying gilded weighted staves. The chassis of our carriage, with C-springs and leather strap suspension and a perch, is as originally built for the Earl of Caledon about 1850 but the present body was substituted in 1891.

2 Landau late 19th century

The carriage shown opposite, known as a canoe-type landau, is an example of one of the most popular and useful of the fashionable carriages of the last century. Since they could be used open or closed they could take the place of the brougham (closed) and the victoria (open) when some economy was necessary. Having been originally introduced in the 18th century as rather stately perch-type dress carriages from the town of Landau in Germany, they gradually became much lighter in construction and the perch was replaced by elliptic springs. Much attention was given to improving them and they remained popular to the end of the century among those who employed a coachman. Landaus normally seated four and were drawn by two horses, but towards the end of the period a humbler one-horse variety was introduced, either for private use or, occasionally, for hire.

3 Clarence late 19th century

In 1838 Lord Chancellor Brougham designed a closed one-horse carriage which was so much smaller and more convenient for town use than the comparatively ponderous vehicles then available that it led to a complete change in carriage design both in this country and on the Continent. It became known as a 'brougham'.

The type of carriage illustrated here, named after the Duke of Clarence, appeared a few years later and was described as being midway between a brougham and a coach. The four-wheeled London cab, the familiar 'Growler', which in its original form had inspired the design of the brougham, was known officially as a Clarence cab.

The Clarence carriage shown here is arranged as an ambulance and in this form it was commonly used in Scotland until the advent of the motor vehicle. It can accommodate two or three patients and an attendant. A special rear door is provided to give easy entry to a stretcher and the inside is lined with polished wood instead of the normal upholstery so as to be easily washed with disinfectant. The rear axle has three-quarter elliptic springs connected by a transverse spring, while the forecarriage has full elliptic springs. The wheels are fitted with pneumatic tyres, although for use as an ordinary carriage they would have had metal or solid rubber tyres.

4 Closed Carriage mid-19th century

Many different types of small closed relatively inexpensive carriage were in use by the professional classes by the middle of the 19th century. This particular example, which was owned by a doctor, was built in Exeter. It is somewhat unusual, not only in the shape of the body but in the fact that the passengers sit facing the back. It is drawn by one horse and driven by a coachman from a box seat at the front. There is a trap door in the roof so that the occupants can communicate with him. A hand brake operates on the rear wheels and elliptic springs are fitted on the front and rear axles.

The solid rubber tyres were probably fitted later, when they came into general use near the end of the century. They were tried first in 1845 but did not become popular until 1886 when a rim with a hollow iron channel was introduced to hold the rubber more firmly in position. In 1892 Kingston Welch designed pneumatic tyres for carriages but they were not a success. They were too prone to distributing mud, too subject to punctures and too difficult to repair.

5 Continental Chaise mid-18th century

Men of rank and fashion took up driving as a sport during the latter part of the 18th century and the following pages show some of the light fast two-wheeled owner-driven vehicles which gradually evolved for town or country use.

The decorative chaise shown opposite is of a type which would have been used by the young aristocracy for driving in towns and on country estates. It has something of 18th century elegance, with its carved and gilded framing, shafts and wheel spokes and its painted panelling. The body is suspended on leather braces and the tyres are iron.

The line drawing below shows a much simpler, more solid and sporting type of chaise. This was used in England about 1765 and demonstrates the early tendency among two-wheelers here to be built with the centre of gravity too far ahead of the axle. Although this tended to prevent backward tipping it proved to put too much weight on the horse.

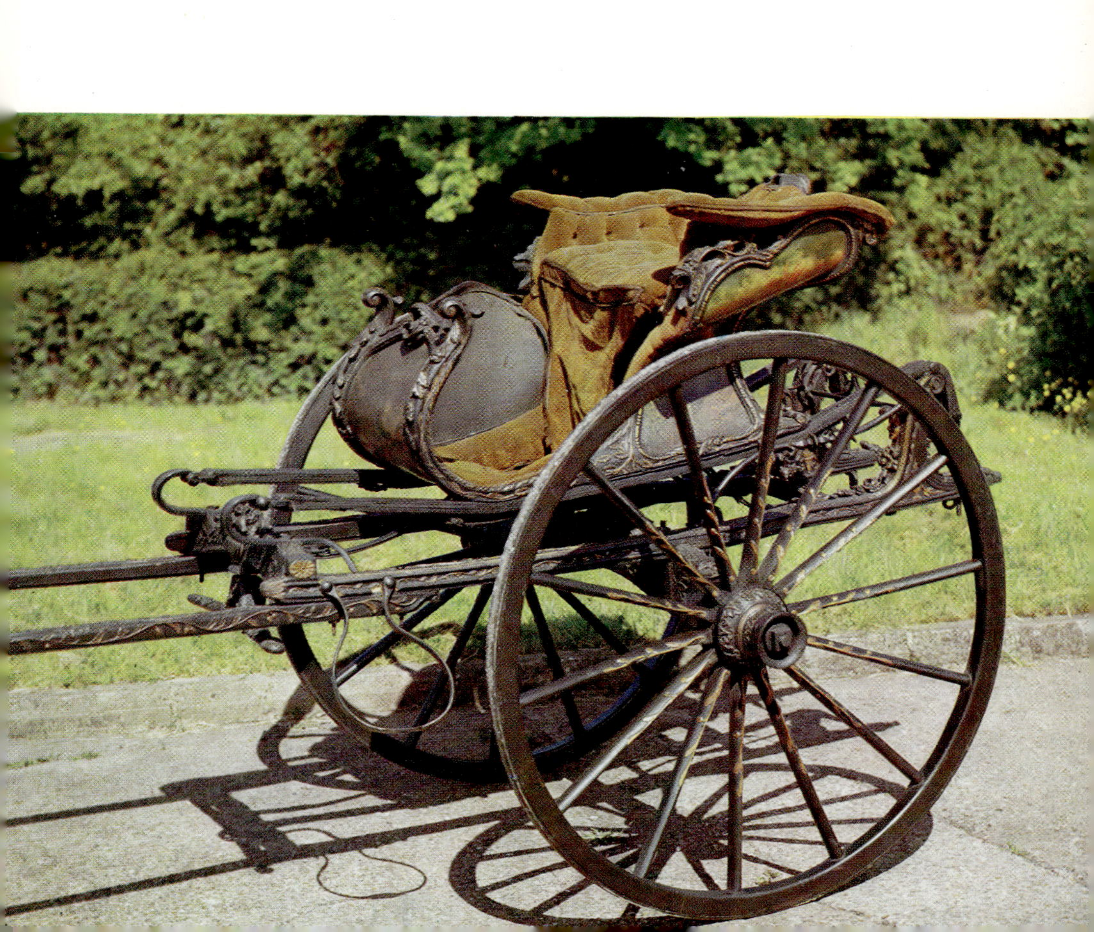

6 Gig mid-19th century

The dangerously high early gigs introduced in the later part of the 18th century were gradually replaced by somewhat safer lower types. Our model shows a typical one-horse gig of the fashionable world of the mid-19th century, by which time it had become one of the most popular of the two-wheeled owner-driven vehicles.
The wheels have metal tyres and the extremely light body is supported on elliptic springs. In actual use there would almost certainly have been a well-padded cushion fitted to the seat. The leather guards fixed at each side, above the lamps, were to protect the occupants from the mud which would have been thrown up by the wheels. Gigs normally carried two people, although two of the earlier types, the Cocking Cart and the Suicide Gig, had a groom's seat behind and the Sulky, used for racing, seated only the driver.

7 Painting of Gig mid-19th century

Carriages were a favourite subject for artists in the 19th century and this oil painting illustrates a well-turned-out gig of the period, with its typically smart high-stepping horse. It was considered elegant for the horse's tail to be docked.

These light vehicles were used for short journeys, for pleasure or for sporting driving in town or country and could be drawn by one horse or by two driven tandem—i.e. one in front of the other. They lent themselves to fast driving but were somewhat unstable, being rather easily overturned, prone to skidding on wet roads and having no brakes.

8 Curricle mid-19th century

The curricle (shown opposite) and the cabriolet (see drawing below) were the aristocrats of the two-wheeled owner-driven carriages. The curricle, which first appeared in the 18th century, was steadier and more comfortable than the early high gigs and became very popular. It was the only English two-wheeled carriage which could take a pair of horses abreast and it seated two, with a groom's seat at the rear. It was not only elegant, with an easy motion, but it was also easy on the horses and was therefore sometimes used for longer journeys as well as for town and park use.

Early in the 19th century it was superseded to some extent as a carriage of fashion for town use by the one-horse cabriolet, imported from France. This had a platform at the back on which stood the groom—a diminutive man or a boy called a 'tiger'—and the owners prided themselves on the smartness and efficiency of both the vehicle and the 'tiger'. From cabriolet we derive the word 'cab' for a hired vehicle, this being the name given to the first successful two-wheeled carriages for hire in London, the hackney cabriolets.

9 Hooded Buggy late 19th century

This type of vehicle was an adaptation of a gig with a hood added—in fact it was sometimes simply called a hooded gig. It was rather more robust than a fashionable gig, but was still a light, speedy one-horse carriage. Although never particularly popular, it was in use until the end of the 19th century and Lord Londsale used a buggy in an epic race in 1891.

In our example the springing is rather unusual, consisting of semi-elliptic springs between axle and shafts and again between shafts and body: also the traces are not hooked directly to the body but to the cross bar.

In America the name 'buggy' was applied to a variety of wagons and carriages, four-wheeled or two-wheeled, although in general it meant a light, fast, horse-drawn vehicle.

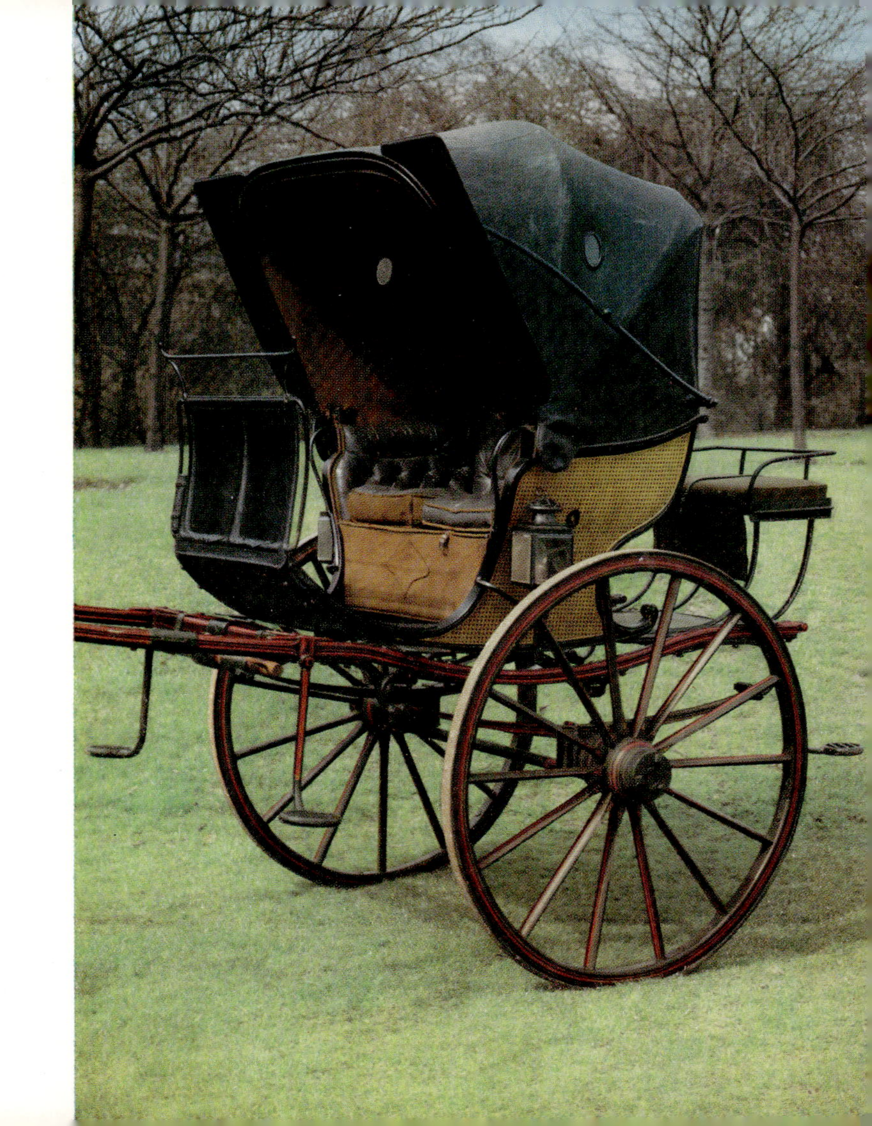

10 Demi-Mail Phaeton late 19th century

In the late 18th century 'phaeton' was a name for a large variety of four-wheeled owner-driven carriages. At first they were built rather high (see drawing below) and were unstable; but they were popular with young sporting bloods, including George IV when Prince of Wales. Lower, more practical types soon appeared and the following pages show some typical examples.
The massive mail phaeton, which first appeared about 1828, was built on a perch like a Mail Coach and was driven exclusively by men. Drawn by two horses, it carried two people, with one or two attendants at the rear and, having plenty of accommodation, could be used for general country purposes or for long journeys.
The demi-mail phaeton, of which our example is shown opposite, was a lighter version introduced a few years later and suitable for town use. It has no perch and the body is mounted on elliptic springing.

THE STATE

11 Stanhope Phaeton 1865

Phaeton carriages of all kinds were used, in town and country, throughout the 19th century and represented a large proportion of England's carriages. The Stanhope is one of the many varieties and was introduced about 1830 to the design of the Hon. Fitzroy Stanhope. It has elliptic springing and illustrates the increasing tendency towards lighter vehicles. It could be drawn by one horse, as our model indicates, and was regarded as a town carriage for gentlemen. There would normally be one or two attendants seated at the back.

The method of attaching the shafts to the carriage by open futchells (which can be seen in the photograph) was developed so that the shafts could be easily detached to save stabling space.

12 Queen Victoria's Pony Bath Chair 1893

This light pony phaeton was built by Cheverton in the Isle of Wight for the use of Queen Victoria when she was old and infirm and it was made low so that entrance was easy.
Suspension is by fully elliptic springs on both axles and open futchells are used with shafts for a pony, which was led by a footman. Behind the hood is a handle which another footman walking behind could use to actuate the brakes on the rear wheels. The mudguards are of leather and there is a leather apron.

13 Dog Cart late 19th century

The next few pages deal with the more utilitarian type of private carriage, chiefly open traps or brakes and used mainly in the country. Increasing in number and variety as roads improved during the 19th century, some even survived until well into the present century in quiet country districts.

Among the humblest but perhaps the most useful of them were the open traps. The type known as a dog cart, of which our example is shown opposite, was introduced early in the 19th century, for carrying guns and dogs for shooting parties and was originally built high. It had a large deep boot with slats in the sides and could take 4 passengers back to back. The two-wheeled type was usually drawn by one horse but occasionally by two driven tandem (see drawing below) and the four-wheeled type could take a pair. After the introduction of a modified design which was lower and more comfortable, the dog cart became indispensable, particularly in the country, appearing in many forms and sizes and being used for an endless variety of duties. Our vehicle seats five in addition to the driver and has fully elliptic springs on both axles.

14 Eridge Cart late 19th century

This rather unusual pony phaeton came originally from Eridge Castle on the Abergavenny estate in Kent and is believed to have been designed by Lord Abergavenny.
Although the undercarriage is similar to that of a pony phaeton, the seating has more resemblance to that of a dog cart, with provision for two facing forward and two facing back. There are fully elliptic springs on both axles and the front wheels are placed well ahead of the body of the vehicle to facilitate turning easily in confined spaces. There is a hand brake operating on the rear wheels.

15 Governess Cart early 20th century

This type of small pony trap or tub cart, originally designed to be used by women, children and elderly people, was low to the ground, safe and easy to get into, light and easy to maintain and was in general use over a long period. It was driven from the offside seat. The driver, as well as the passengers, sat sideways to the centre, so a quiet, easily controlled pony was preferable. The recess in the cushion helped to some extent, by providing space for the driver's legs and enabling him to turn slightly forward. The later types, such as that illustrated here, were well balanced, being so designed that with careful loading the weight was taken directly above the axle and the load on the shafts was minimal.

16 Charabanc c.1890

The charabanc, our model of which is shown in the photograph, was a coachman-driven utility vehicle, owned mainly by large establishments in the country and drawn by four horses. It was introduced here from France about 1850.

The wagonette (see drawing below) appeared about the same time. The Prince Consort ordered one and it became a favourite vehicle with Queen Victoria and her children. It was soon very popular all over the country, taking many forms, from the relatively luxurious to the humbler vehicles used by farmers, which were sometimes made with removeable seats for the transport of goods. It could be used open or closed and could be drawn by one horse or two.

The charabanc carried more people than the wagonette and was useful to transport large shooting parties or staff and equipment. Later it was used for public sight-seeing until eventually superseded by the modern motorized version.

CHAPMAN & SONS.

17 Hansom Cab late 19th century

In town there had been vehicles for hire throughout the 17th and 18th centuries in the form of hackney carriages, supplanted in the 18th century by two-wheeled and four-wheeled cabs and the passenger omnibus. For longer journeys, from about the beginning of the 18th century, a posting chariot could be hired, with relays of horses and drivers. For the majority, however, there was the stage coach (which had come into use by 1640) or, from the end of the 18th century, the Mail Coach.

Our model of a hansom cab is not entirely representative of the normal hired vehicle, but is of particular interest as an example of a privately owned cab, having been based on one owned by Lord Rosebery. Like the hired variety it provides seating for two passengers, who can speak to the driver through a trap door in the roof. The flap doors are manipulated by the driver by means of a lever and chains and there would have been windows above the doors, which could be opened by the passengers. Public hansom cabs had provision for carrying luggage on the roof.

18 Knifeboard Omnibus 1855

The omnibus was introduced into England in 1829 by John Shillibeer, an English coachbuilder living in Paris. Its success was immediate in spite of opposition from the hackney coachmen.

The drawing below shows one which was in use in London about 1840.

The model represents a type of London omnibus of 1855 and has the low roof seat known as a 'knifeboard', which was added so that more passengers could be carried. The conductor stood on a small platform beside the rear door. There were three sliding windows and two louvre ventilating panels on each side and a sliding louvre panel in the door. Cushions were provided on all seats. There were elliptic springs on the forecarriage and one transverse and two side semi-elliptic springs at the rear. The vehicle was drawn by two horses, carried 12 passengers inside and 15 outside, and could travel at about 6 to 8 mph.

KENNINGTON
CITY
CLAPHAM
CITY

19 Post-chaise late 18th century

During the greater part of the 18th century and the early part of the 19th the hired post-chaise afforded a speedier and more comfortable method of travelling long distances, for those who could afford it, than the stage coach or the mail coach.

The model shows one of the 'yellow bounders' as they were known —they were nearly always gentlemen's discarded private travelling chariots painted yellow. They seated two passengers. were drawn by four hourses, driven postillion, and average speeds up to 10 mph might be maintained according to the prevailing road conditions. Relays of horses were available from establishments along the routes, known as 'posts'. The same posts would also provide fresh postillion drivers and horses for those who could afford their own private travelling carriages.

20 Mail Coach 1827

Shown opposite is the original London to York Royal Mail Coach which was in use from 1827 to about 1860 and is now in the Museum's Collection. It could take eight passengers altogether, the first class inside, the second class on top and its average speed was 10 mph. Only the two small front lamps are shown but at the side can be seen one of the holders for the additional two very large lamps which are now fitted to the coach in the gallery.
The Royal insignia painted on the sides were originally those of William IV but were later changed to those of Queen Victoria. The mail was kept in the rear compartment, on top of which is the seat for the guard and a projecting case for his blunderbuss.
For shorter journeys, of perhaps 20 or 30 miles, passengers could travel on what were called 'short stage coaches' (see drawing below), which did not carry mail and could be drawn by two horses instead of four.

YORK
LONDON
ROYAL
MAIL

Science Museum illustrated booklets

Other titles in this series:

Timekeepers Clocks, watches, sundials, sand-glasses

Ship Models Part 1 : From earliest times to 1700 AD

Ship Models Part 2 : Sailing ships from 1700 AD

Ship Models Part 3 : British Small Craft

Ship Models Part 4 : Foreign Small Craft

Railways Part 1 : To the end of the 19th century

Railways Part 2 : the 20th century

Chemistry Part 1 : Chemical Laboratories and Apparatus to 1850

Chemistry Part 2 : Chemical Laboratories and Apparatus from 1850

Chemistry Part 3 : Atoms, Elements and Molecules

Chemistry Part 4 : Making Chemicals

Aeronautics Part 1 : Early Flying up to the Reims meeting

Aeronautics Part 2 : Flying since 1913

Power to Fly Aircraft Propulsion

Aeronatica Objets d'Art, Prints, Air Mail

Lighting Part 1 : Early oil lamps, candles

Lighting Part 2 : Gas, mineral oil, electricity

Lighting Part 3 : other than in the home

Making Fire Wood friction, tinder boxes, matches

Cameras Photographs and Accessories

Agriculture Hand tools to Mechanization

Fire Engines and other fire-fighting appliances

Astronomy Globes, Orreries and other Models

Surveying Instruments and Methods

Physics for Princes The George III Collection

Motor Cars up to 1930

Steamships Merchant ships to 1880

British Warships 1845–1945

Published by
Her Majesty's Stationery Office
and obtainable from the
Government Bookshops listed
on cover page iv (post orders
to PO Box 569 London SE1)

7s each (by post 7s 4d)

Printed in England for
Her Majesty's Stationery Office
by W. Heffer & Sons Ltd.
Cambridge

Dd. 147312 K 160